;ontents

'he road user on f

'he road user on wheels

This Code, between pages 6 and 45, is issued with the Authority of Parliament (Resolutions passed November 1977)

A failure on the part of a person to observe a provision of the Highway Code shall not of itself render that person liable to criminal proceedings of any kind, but any such failure may in any proceedings (whether civil or criminal, and including proceedings under this Act, the Road Traffic Regulation Act 1967 or the Public Passenger Vehicles Act 1981) be relied upon by any party to the proceedings as tending to establish or to negative any liability which is in question in those proceedings.

Road Traffic Act 1972, Section 37

The road user on foot

GENERAL

1 Where there is a pavement or footpath, use it. Do not walk next to the kerb with your back to the traffic. Look both ways before you step into the road.

2 Where there is no footpath, walk on the right-hand side of the road — it is safer to walk on the side facing oncoming traffic. Keep as close as possible to the side of the road, and do not walk more than two side by side. Take care at right-hand bends. Keep one behind the other if possible, particularly in heavy traffic or in poor light.

3 Do not allow children — up to age five at least — out alone on the road. Go with them, walk between them and the traffic and always keep tight hold of their hands; if you cannot do this, then use reins or secure them firmly in a pushchair. Do not let them run into the road.

4 Always wear or carry something white or light-coloured or reflective in the dark or in poor light. This is especially important on roads without footpaths. (Reflective material can be seen in headlights at up to three times the distance when compared with ordinary clothes; fluorescent material is highly conspicuous in daylight and at dusk but is of little use in the dark.)

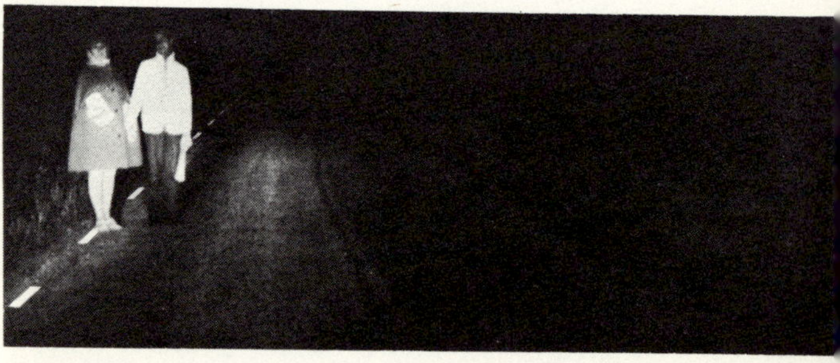

5 A group of people marching on the road should keep to the left. There should be look-outs in front and at the back wearing reflective clothing at night and fluorescent clothing by day. At night the look-out in front should carry a white light, and the one at the back should carry a bright red light visible from the rear. Additional lights should be carried and reflective clothing worn by the outside rank of long columns.

6 You must not walk on motorways or their slip roads to thumb lifts or for any other purpose (see Rule 145).

CROSSING THE ROAD

7 The Green Cross Code

The Green Cross Code is a guide for all pedestrians. However, children need to be taught how to use it and should not be allowed out alone until they can understand and apply it. The age at which they can do this will vary: for instance, many children under seven cannot fully understand and apply those parts of the code requiring judgement of the speed and distance of approaching vehicles. Teaching children the code and the age at which parents allow them to go out and cross roads by themselves must therefore be suited to each individual child.

1 First find a safe place to cross, then stop.

It is safer to cross at subways, footbridges, islands, Zebra and Pelican crossings, traffic lights or where there is a policeman, a "lollipop" man or a traffic warden. If you can't find any good crossing places like these, choose a place where you can see clearly along the roads in all directions. Don't try to cross between parked cars. Move to a clear space and always give drivers a chance to see you clearly.

2 Stand on the pavement near the kerb.

Don't stand too near the edge of the pavement. Stop a little way back from the kerb — where you'll be away from traffic, but where you can still see if anything is coming. If there is no pavement, stand back from the edge of the road but where you can still see traffic coming.

3 Look all round for traffic and listen.

Traffic may be coming from all directions, so take care to look along every road. And listen, too, because you can sometimes hear traffic before you can see it.

4 If traffic is coming, let it pass. Look all round again.

If there's any traffic near, let it go past. Then look round again and listen to make sure no other traffic is coming.

5 When there is no traffic near, walk straight across the road.

When there is no traffic near it's safe to cross. If there is something in the distance do not cross unless you're *certain* there's plenty of time. Remember, even if traffic is a long way off, it may be coming very fast. When it's safe, walk straight across the road — don't run.

6 Keep looking and listening for traffic while you cross.

Once you're in the road, keep looking and listening in case you didn't see some traffic — or in case other traffic suddenly appears.

Crossing where there is an island in the road

8 Use the Green Cross Code (see Rule 7) to cross to the island. Stop there and use the Code again to cross the second half of the road.

Crossing at a junction

9 When you cross at a road junction look out for traffic turning the corner, especially from behind you.

Crossing at a Zebra crossing (see illustration page 57)

10 If there is a Zebra crossing near, always use it. It is very dangerous to cross the road a short distance away from a Zebra crossing. The most dangerous area is usually marked with zigzag lines — don't cross on them.

11 Always give drivers plenty of time to see you and to slow down and stop before you start to cross. If necessary put one foot on the crossing. Until you have stepped on to a Zebra crossing, the traffic does not have to stop for you. Vehicles need more time to stop if the road is slippery because of rain or ice.

12 When the traffic has stopped, walk across but keep looking both ways and listening in case a driver has not seen you and tries to overtake a vehicle which has stopped.

13 If there is an island in the middle of a crossing, pause on it and obey Rules 11 and 12 before crossing the second half of the road — it is a separate crossing.

Crossing at a Pelican crossing

14 Always use a Pelican crossing if one is available, even if you have to walk a bit further to do so. At this type of crossing the traffic is controlled by traffic lights and there is a light signal to tell pedestrians when to cross.

AT TRAFFIC LIGHTS AND PELICAN CROSSINGS

AT PELICANS ONLY

When the red man signal shows, don't cross. Press the button on the box and wait. The lights will soon change and a steady green man signal will appear; you may now cross with care. (At some Pelicans there is also a bleeping sound to tell blind people when the steady green man signal is showing.)

After a short time, the green man signal will begin to flash. This means that the lights will soon change again. You should not start to cross but if you have started already you will have time to finish crossing safely.

15 A Pelican which goes straight across the road is one crossing even if there is a central refuge. But if the crossing is staggered (separate crossings one each side of the refuge but not in a straight line) you must press the button on the refuge to get the green man signal for the second crossing.

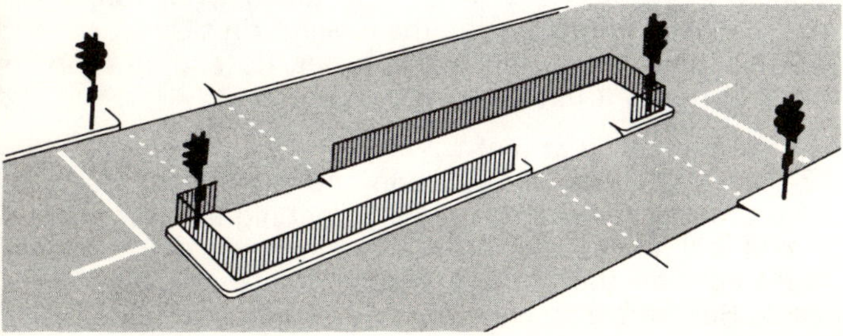

Crossing at traffic lights

16 If there are pedestrian signals, obey them. If not, watch both the lights and the traffic and do not cross when the lights allow traffic to go forward, even if you think you have time. When the lights are red remember to look out for turning traffic, and remember also that some traffic lights allow traffic to proceed in some lanes while other lanes are stopped.

Crossings controlled by police, traffic wardens or school crossing patrols

17 When a policeman, traffic warden or school crossing patrol is controlling the traffic, cross the road only when the officer has signalled you to do so. Always cross in front of him.

Guard rails

18 Where there are guard rails, cross the road only at the gaps provided for pedestrians. Do not climb over the guard rails or walk outside them.

Crossing one-way streets

19 Use the Green Cross Code. Check which way the traffic is going and remember that in one-way streets there will usually be more than one lane of traffic going in the same direction. Do not cross until it is safe to cross all the lanes of traffic.

Crossing bus lanes

20 Use the Green Cross Code. Vehicles in bus lanes may be going faster than traffic in other lanes and in the opposite direction.

Parked vehicles

21 Try not to cross the road where there are parked vehicles. If you have to cross between parked vehicles, stop at the outside edge of the vehicles where you can be seen by drivers and where you can look all round for traffic. Then continue using the Green Cross Code. Do not stand behind any vehicle which has its engine running because it might back into you.

Crossing the road at night

22 Use the Green Cross Code and remember that it is more difficult for drivers to see you at night. If there is no convenient pedestrian crossing or island, cross near a street light so that drivers will be able to see you more easily. Remember to wear or carry something reflective or light-coloured so that you can be seen more easily in the dark.

EMERGENCY VEHICLES

23 Keep out of the road if you see or hear ambulances, fire engines or police or other emergency vehicles with their blue lamps flashing, or their bells, two-tone horns or sirens sounding.

GETTING ON OR OFF A BUS

24 Do not get on or off a bus unless it is standing at a bus stop. If you want to get on a bus at a request stop, give a clear signal for the bus to stop and do not try to get on until it has done so. Never cross behind or in front of a bus. Wait until it has moved off and you have a clear view of the road in both directions.

RAILWAY LEVEL CROSSINGS

25 Take particular care at level crossings (see Rules 176–185).

The road user on wheels

(The Department of Transport's manual *Driving* deals with the following points in greater detail.)

GENERAL

26 Keep your vehicle in good condition. Pay particular attention to lights, brakes, steering, tyres (including spare), seat belts, demisters, windscreen wipers and washers. Keep windscreens, windows, lights, direction indicators, reflectors, mirrors and number plates clean and clear. Do not drive with a defective or unsuitable exhaust system. If L-plates have been fitted, remove (or cover) them when the vehicle is not being used for driving instruction or practice.

27 You must ensure that any loads carried or towed are secure.

28 If you are involved in an accident, wearing a seat belt halves the risk of death or serious injury. Drivers and front seat passengers in most vehicles must wear a seat belt. Those exempt from the law include holders of a medical exemption certificate, drivers carrying out a manoeuvre which includes reversing, and those engaged in making local rounds of deliveries and collections in a vehicle constructed or adapted for that purpose. It is your responsibility to wear the belt unless exempt. Make sure your seat belts are properly adjusted.

Children under 14 must be restrained if they are travelling in the front. For a child under one year this means an approved child restraint designed for a child of that age and weight. A child under 14 but over one may wear any approved child restraint or an adult seat belt. It is the driver's responsibility to ensure that a child under 14 is suitably restrained. The safest place for a young child to travel is restrained in the rear seat of the car, wearing a properly fitted child safety harness which bears an approved marking to show that it has been tested and approved to the appropriate standard. Ensure that child safety locks on doors, where fitted, are secured when children are being carried.

29 When on a motorcycle, scooter or moped, you must wear a safety helmet of approved design. You should also wear light-coloured or reflective and fluorescent clothing. (See Rule 4.)

30 Do not start a journey if you feel tired. Never drive after drinking alcohol or while under the influence of drugs. If you have to take drugs in the course of medical treatment consult your doctor about their possible effect on your driving ability. Do not drive if you are unwell.

31 If you need spectacles to meet the official eyesight standard, wear them. It is an offence to drive with uncorrected defective vision.

32 Do not use tinted optical equipment of any kind (sunglasses, night driving spectacles, ski goggles, tinted helmet visors) at night or in conditions of poor visibility.

33 Tinted glass does not help your vision. Do not use spray-on or other tinting materials for windows and windscreens.

SIGNS

34 Know your traffic signs and road markings (see pages 50 – 55) and act on them always.

SIGNALS

35 Give signals if they would help or warn other road users. Give only the correct signals — those illustrated on pages 48 and 49. Give them clearly and in good time. Always be sure that your direction indicator signals are cancelled after a manoeuvre.

36 Watch out for other drivers' signals and take any necessary action promptly.

37 You must obey signals given by police officers and traffic wardens directing traffic (see page 47) or signs displayed by school crossing patrols.

MOVING OFF

38 Before moving off, always use your mirror; but look round as well for a final check. Signal if necessary before moving out; move off only when you can do so safely without making other road users change speed or direction.

DRIVING ALONG

39 Keep to the left, except when road signs or markings indicate otherwise or when you intend to overtake, or turn right, or when you have to pass stationary vehicles or pedestrians in the road. Allow others to overtake you if they want to. You must not drive on a footpath or pavement by the side of the road.

40 Use your mirror often, so that you know what is behind you.

41 On narrow or winding roads, or where there is a lot of oncoming traffic, drivers of large or slow-moving vehicles should be prepared to pull in, and slow down or stop, as soon as there is a suitable opportunity to do so, to give faster vehicles a chance to overtake.

42 Well before you overtake, or turn left, or turn right, or slow down, or stop, use your mirror (motorcyclists should always glance behind, even if they have mirrors fitted); then give the appropriate signal if necessary.

Remember the routine: **Mirror — Signal — Manoeuvre.**

43 Always keep a special look out for cycles and motorcycles, particularly when overtaking or turning. Bear in mind that two-wheelers are much less easy to see than larger vehicles and that their riders have the same rights to consideration as other road users and are more vulnerable. Drivers (especially of long vehicles or of vehicles towing trailers) should leave plenty of room for pedal cyclists in particular.

44 Driving for long distances may make you feel sleepy. To help prevent this, make sure there is plenty of fresh air in your vehicle. If you become tired on a journey, stop and rest at a suitable parking place.

45 You must obey the speed limits for the road and for your vehicle. Remember that, except on motorways, there is a 30 mph speed limit on all roads where there are street lights unless signs show otherwise. Bear in mind that any speed limit is a maximum. It does not mean that it is safe to drive at that speed — always take into account all the conditions at the time.

46 Never drive so fast that you cannot stop well within the distance you can see to be clear. Go much more slowly if the road is wet or icy or if there is fog. Do not brake sharply except in an emergency.

47 Leave enough space between you and the vehicle in front so that you can pull up safely if it slows down or stops suddenly. The safe rule is never to get closer than the overall stopping distance shown below. But on the open road, in good conditions, a gap of one yard for each mph of your speed or a two-second time gap may be enough. This will also leave space for an overtaking vehicle to pull in. On wet or icy roads the gap should be at least doubled. Drop back if an overtaking vehicle pulls into the gap in front of you.

Shortest stopping distances — in feet

mph	Thinking distance	Braking distance	Overall stopping distance
20	20	20	40
30	30	45	75
40	40	80	120
50	50	125	175
60	60	180	240
70	70	245	315

(See diagram on back cover)

On a dry road, a good car with good brakes and tyres and an alert driver will stop in the distances shown. Remember these are shortest stopping distances. Stopping distances increase greatly with wet and slippery roads, poor brakes and tyres, and tired drivers.

48 Make way for ambulances, fire engines or police and other emergency vehicles when their blue lamps are flashing or their bells, two-tone horns or sirens are sounding.

49 In towns give way to buses indicating an intention to move out from bus stops if you can do so safely.

DRIVING IN FOG

50 The Fog Code

1 Slow down; keep a safe distance. You should always be able to pull up within your range of vision.

2 Don't hang on to someone else's tail lights; it gives a false sense of security.

3 Watch your speed; you may be going much faster than you think. Do not speed up to get away from a vehicle which is too close behind you.

4 Remember that if you are driving a heavy vehicle it may take longer to pull up than the vehicle ahead.

5 Warning signals are there to help and protect; observe them.

6 See and be seen — you must use your headlamps or front fog lamps. Use your windscreen wipers.

7 Check and clean windscreens, lights, reflectors and windows whenever you can.

8 If you must drive in fog, allow more time for your journey.

THE SAFETY OF PEDESTRIANS

51 Drive carefully and slowly when pedestrians are about, particularly in crowded shopping streets, when you see a bus stopped, or near a parked milk float or mobile shop. Watch out for pedestrians emerging suddenly, for example from behind parked or stopped vehicles. Remember, pedestrians may have to cross roads where there are no crossings; show them consideration.

52 Two out of three pedestrians killed or seriously injured are either under 15 or over 60. The young and the elderly may not judge speeds very well and may step into the road when you do not expect them. Watch out for blind people who may be carrying white sticks (white with two red reflectorised bands for deaf/blind people) or using guide dogs, and for the disabled or infirm. Give them plenty of time to cross the road. Remember that deaf people may not hear your vehicle approaching.

53 Drive slowly near schools and look out for children getting on or off buses. Stop when signalled to do so by a school crossing patrol showing a "STOP — CHILDREN" sign. In places of particular danger, there may be a flashing amber signal below the advance sign which warns of a school crossing patrol operating ahead.

54 Be careful near a parked ice-cream van — children are more interested in ice-cream than in traffic.

55 When coming to a Zebra crossing, keep a look out for pedestrians waiting to cross (particularly children, the elderly, the infirm and people with prams) and be ready to slow down or stop to let them cross. When anyone has stepped on to a crossing, you must give way. Signal to other drivers that you mean to slow down or stop. Allow more time for stopping on wet or icy roads. Do not signal pedestrians to cross: another vehicle may be approaching.

56 In the area marked by zigzag lines on the approach to a Zebra crossing, you must not overtake the moving motor vehicle nearest the crossing, or the leading vehicle which has stopped to give way to a pedestrian on the crossing. Even when there are no zigzags never overtake just before a Zebra crossing.

57 In traffic queues, leave pedestrian crossings clear.

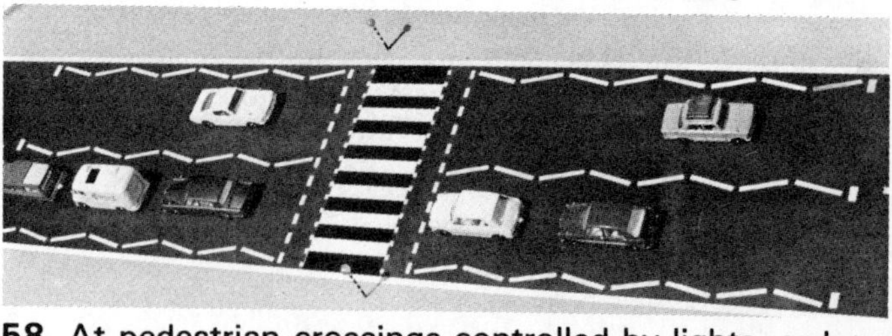

58 At pedestrian crossings controlled by lights, or by a police officer or traffic warden, give way to pedestrians who are still crossing when the signal allows vehicles to move.

AT PELICANS

Give way to pedestrians

59 At Pelican crossings the signals have the same meaning as traffic lights except that a flashing amber signal will follow the red "STOP" signal. When the amber light is flashing you must give way to any pedestrians on the crossing; otherwise you may proceed. A straight Pelican is one crossing even when there is a central refuge and you must wait for people crossing from the further side of the refuge. Don't harass pedestrians — for instance, by revving your engine.

60 When turning at a road junction, give way to pedestrians who are crossing the road into which you are turning.

61 When entering or emerging from property bordering on a road, give way to pedestrians as well as to traffic on the road.

62 Be careful when there are pedestrians, processions or other marching groups in the road, particularly where there is no footpath. Give them plenty of room. Be especially careful on a left-hand bend and keep your speed down.

ANIMALS

63 Go slowly when driving past animals. Give them plenty of room and be ready to stop if necessary. Do not frighten the animals by sounding your horn or revving your engine. Watch out for animals being led on your side of the road and be especially careful at a left-hand bend.

SINGLE-TRACK ROADS

64 Some roads (often called single-track roads) are only wide enough for vehicles to move in one direction at a time. They have special passing places. When you see a vehicle coming towards you, or the driver behind you wants to overtake and the passing place is on your side, pull in; if it is on the other side, wait opposite it. Give way to vehicles coming uphill whenever possible. Do not park in passing places.

LINES AND LANES ALONG THE ROAD

65 A single broken line, with long markings and short gaps, in the middle of the road is a hazard warning line. Do not cross it unless you can see that the road well ahead is clear.

66 Where there are double white lines along the road and the line nearer to you is solid, you must not cross or straddle it except when you need to get in and out of premises or a side road, or when you are ordered to cross the lines by a policeman or traffic warden, or when you have to avoid a stationary obstruction.

67 Where there are double white lines along the road and the line nearer to you is broken, you may cross the lines to overtake if you can do so safely and before reaching a solid white line on your side. It is up to you to be sure it is safe.

68 Areas of white diagonal stripes or white chevrons painted on the road are to separate traffic streams liable to be a danger to each other or to protect traffic turning right. Do not drive over these areas if you can avoid doing so. Where the chevron has a solid white edge line you must not enter the area except in an emergency.

69 Keep between the traffic lane markings — the short broken white lines which divide the road into lanes. Keep in the left-hand lane unless you are going to overtake, turn right or pass parked vehicles. Coloured reflecting road studs may be used with white lines — white studs mark the lanes or centre of the road, while the edge of the carriageway may have red studs on the left-hand side and amber by the central reservation of dual carriageways. Green studs may be used across lay-bys and side roads.

70 Do not move unnecessarily from lane to lane. If you need to move into another lane, first use your mirror. If it is safe to move over, signal before doing so. Make sure you will not force another driver to swerve or slow down.

71 When coming to junctions, be guided by any lane indication arrows on the road or on signs.

72 In a traffic hold-up, do not try to "jump the queue" by cutting into another lane or by overtaking the vehicles waiting in front of you.

73 If a single-carriageway (i.e., undivided) road has three lanes, use the middle lane only for overtaking and turning right. Remember that you have no more right to use the middle lane than a driver coming from the opposite direction. Do not use the right-hand lane.

74 If a single-carriageway road has four or more lanes do not use the lanes on the right-hand half of the road unless signs and markings indicate that you may do so.

75 On a three-lane dual carriageway you may stay in the middle lane when there are slower vehicles in the left-hand lane, but you should return to the left-hand lane when you have passed them. The right-hand lane is for overtaking (or for right-turning traffic); if you use it for overtaking move back into the middle lane and then into the left-hand lane as soon as you can, but without cutting in.

76 In one-way streets, choose the correct lane for your exit as soon as you can. Never change lanes suddenly. Unless road markings indicate otherwise, choose the left-hand lane when going to the left, the right-hand lane when going to the right and any convenient lane when going straight on. Remember – other vehicles could be passing on both sides.

77 A bus lane, which is shown by signs and road markings, may operate for 24 hours or for other periods shown by time plates. Details of those vehicles which are permitted to use the special lane are indicated on the signs. Outside the indicated period of operation all vehicles may use the bus lane.

OVERTAKING

78 Do not overtake unless you are sure you can do so without danger to others as well as yourself. Before you start to overtake make sure that the road is clear far enough ahead and behind. Use your mirror and, if you are on a motorcycle, glance behind. Signal before you start to move out. Be particularly careful at dusk, in the dark and in fog or mist, when it is more difficult to judge speed and distance.

Remember: **Mirror — Signal — Manoeuvre.**

79 On fast roads, vehicles may be coming up behind much more quickly than you think. Make sure that the lane you will be moving into is clear for a long way behind.

80 Once you have started to overtake, move quickly past the vehicle you are overtaking and leave it plenty of room. Then move back to the left side of the road as soon as you can, but without cutting in.

81 Overtake only on the right, except:

a when the driver in front has signalled that he intends to turn right and you can overtake him on the left without getting in the way of others and without entering a bus lane during its period of operation;

b when you want to turn left at a junction;

c when traffic is moving slowly in queues and vehicles in a lane on the right are moving more slowly than you are;

d in one-way streets (but not dual carriageways) where vehicles may pass on either side.

82 When traffic is moving as described at Rule 81*c*, you may move to a lane on your left only in order to turn left or to park. Do not change lanes to the left in order to overtake.

83 Do not increase your speed while being overtaken. Slow down, if necessary, to let the overtaking vehicle pass.

84 On the two-lane road, give way to vehicles coming

towards you before you pass parked vehicles or other obstructions on the left-hand side of the road.

85 You MUST NOT overtake:

— if you would have to cross or straddle double white lines with a solid line nearer to you;

— if you are within the zigzag area on the approach to a Zebra crossing (see Rule 56 for definition);

— after a "No Overtaking" sign and until the end of the restriction.

Do not overtake:

where you cannot see far enough ahead to be sure it is safe to do so, for example, when at or coming to:

— a corner or bend;

— a hump-backed bridge;

— the brow of a hill; or

where you might come into conflict with other road users, for example,

— at a road junction;

— at a level crossing;

— where the road narrows;

— on the approach to any type of pedestrian crossing;

— where it would involve driving over an area marked with diagonal stripes or chevrons.

Do not overtake:

— when to do so would force another vehicle to swerve or slow down.

IF IN DOUBT — DO NOT OVERTAKE.

ROAD JUNCTIONS

86 Approach junctions with great care. Consider your road position and your speed. Drive on only when you are sure it is safe to do so and that you will not block the junction. Watch out for long vehicles which may be turning — left or right — at a junction ahead but which may have to use the whole width of the road in order to make the turn.

87 When waiting to emerge at a junction do not assume that a vehicle approaching from the right which is signalling with its left-hand direction indicator *will* turn left. Wait to make sure.

88 At a junction with double broken white lines across the road (it may also have a "Give Way" sign or an inverted triangle on the carriageway) you must be ready to let traffic on the major road go by first.

89 At junctions with a "STOP" sign and a solid white line across your approach, you must stop at the line. Wait for a safe gap in the traffic before you move off.

90 When crossing a dual carriageway or turning right into one, treat each half as a separate road. Wait in the central dividing strip (the central reservation) until there is a safe gap in the traffic on the second half of the road. However, where the central reservation is too narrow for the length of your vehicle, you should wait in the side road until you can cross the dual carriageway in one movement.

91 Where a junction has a mini-roundabout, it will have a sign (as shown on page 50) placed before the "Give Way" line and Rules 102—107 must be obeyed.

Mini-roundabout

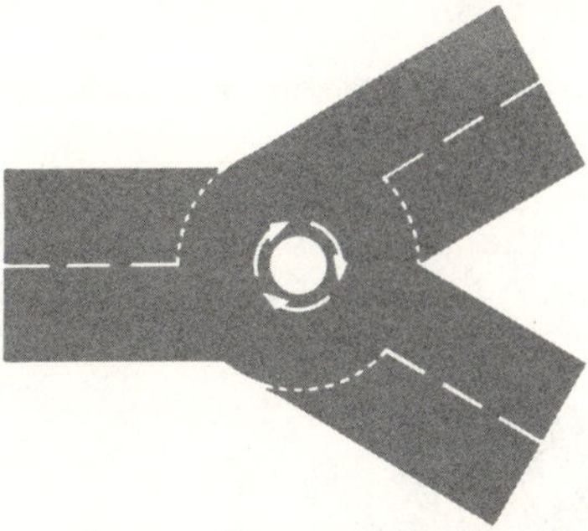

92 Box junctions have criss-cross yellow lines painted on the road. You must not enter the box if your exit road or lane from it is not clear. But you may enter the box when you want to turn right and are prevented from doing so only by oncoming traffic or by vehicles waiting to make a right turn.

Junctions controlled by police or traffic wardens
93 When all traffic is held up by a police officer or traffic warden, you must not filter to the left or right until he signals you to do so.

Junctions controlled by traffic lights
94 Do not go forward when the traffic lights are green unless there is room for you to clear the junction safely.

Never go forward when the red and amber lights are showing together.

95 Where traffic lights have a green arrow filter signal, do not get into the lane where filtering is allowed unless you want to go in the direction shown by the arrow.

96 At junctions controlled by traffic lights, vehicles required to stop must wait behind the solid white ''STOP'' line marked across the approach.

TURNING RIGHT

97 Well before you turn right, use your mirror to make sure you know the position and movement of traffic behind you. When it is safe, give a right turn signal and, as soon as you can do so safely, take up position just left of the middle of the road or in the space marked for right-turning traffic. If you can, leave room for other vehicles to pass on the left. Wait until there is a safe gap between you and any oncoming vehicle; then make the turn, but do not cut the corner. Give way to pedestrians crossing the road into which you are turning. Remember: **Mirror — Signal — Manoeuvre.**

98 When turning right at a junction where there is an opposing vehicle also turning right, drive your vehicle so that you keep it to your right and pass behind it (offside to offside). Check for other traffic on the carriageway you intend to cross before completing the turn. If the layout of the junction or the traffic situation is such that offside-to-offside passing is impractical, or if nearside-to-nearside passing is indicated either by road markings or by the other driver, watch carefully for traffic approaching on the carriageway you intend to cross, which may be masked by the other vehicle.

99 When turning right from a dual carriageway, wait in the opening in the central reservation until it is safe to cross the second half of the road.

TURNING LEFT

100 Well before you turn left, use your mirror and give a left turn signal. Before and after the turn keep as close to the left as safety and the length of your vehicle will allow. Make sure a cyclist or motorcyclist is not coming up from behind on your left. Give way to pedestrians crossing the road into which you are turning.

101 If you intend to turn left across a bus lane on your left, look out for any vehicles, particularly bicycles, that may be in the bus lane.

ROUNDABOUTS

102 When entering a roundabout, give way to any traffic on your immediate right unless road markings indicate otherwise; but keep moving if the way is clear. At some junctions there may be more than one roundabout. At each, apply the normal rules for roundabouts. Keep a special look out for the ''Give Way'' lines.

103 Where there are two lanes at the entrance to a roundabout, unless signs or road markings indicate otherwise:

When turning left:
Approach in the left-hand lane; keep to that lane in the roundabout.

When going forward:
Approach in the left-hand lane; keep to that lane in the roundabout. If conditions dictate (for example, if the left-hand lane is blocked), approach in the right-hand lane; keep to that lane in the roundabout. If the roundabout itself is clear of other traffic, take the most convenient lane through the roundabout.

When turning right:
Approach in the right-hand lane; keep to that lane in the roundabout.

104 When there are more than two lanes at the entrance to a roundabout, unless signs or road markings indicate otherwise, use the clearest convenient lane on approach and through the roundabout suitable for the exit you intend to take.

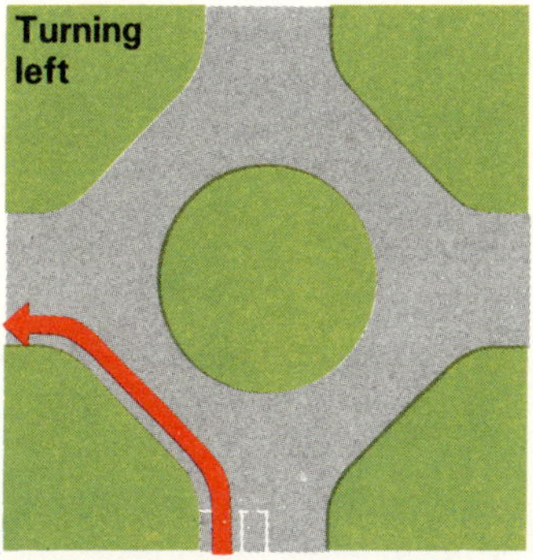

Procedure at roundabouts with a two-lane entrance

The recommended course in each situation is shown by a solid line: where conditions dictate, drivers may follow the course indicated by the broken line.

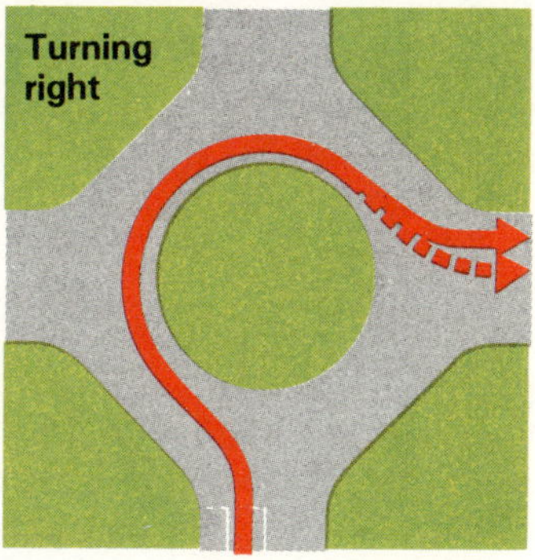

105 When in a roundabout, look out for and show consideration to other vehicles crossing in front of you, especially those intending to leave by the next exit.

106 Signals at roundabouts:

When turning left:
Use the left turn indicator on approach and through the roundabout.

When going forward:
Use the left turn indicator when passing the exit before the one to be taken.

When turning right:
Use the right turn indicator on approach, and maintain this signal until passing the exit before the one to be taken. Then change to the left turn indicator.

107 Watch out for pedal cyclists and for long vehicles which may have to take a different course, both on the approach to and in the roundabout.

REVERSING

108 Before you reverse, make sure that there are no pedestrians — particularly children — or obstructions in the road behind you. Be especially careful about the "blind area" behind you — that is, the part of the road which you cannot see from the driving seat.

109 If you cannot see clearly behind, get someone to guide you when you reverse.

110 Never reverse from a side road into a main road. You must not reverse your vehicle for longer than is necessary.

LAMPS

111 You must:

a make sure that all your lamps are clean, that they work and that your headlamps are properly adjusted — badly adjusted headlamps can dazzle road users and lead to accidents;

b switch on your lamps at lighting-up time;

c use headlamps at night on all roads where there is no street lighting, on roads where the street lamps are more than 200 yards apart and on roads where the street lamps are not alight.

d use your headlamps or front fog lamps at any time when visibility is *seriously* reduced, that is, generally, reduced to a distance of less than 100 metres.

REAR FOG LAMPS

111A You must not use your rear fog lamps unless visibility is *seriously* reduced, that is, generally, reduced to a distance of less than 100 metres. Do not use them simply because it is dark or raining or misty.

112 You should also:

a use headlamps at night on lighted motorways and similar high speed roads;

b use dipped headlamps at night in built-up areas unless the road is well lit;

c always drive so you can stop well within the distance you can see ahead;

d slow down or stop if you are dazzled by approaching headlamps;

e dip your headlamps when meeting other vehicles or road users and before they dazzle the driver of a vehicle travelling in the same direction in front of you.

FLASHING HEADLAMPS

113 The flashing of headlamps has only one meaning — like sounding your horn it lets another road user know you are there. Do not flash your headlamps for any other reason.

USE OF THE HORN

114 When your vehicle is moving use your horn when it is necessary as a warning of your presence to other road users — but never use it as a rebuke. You must not use your horn when your vehicle is moving between the hours of 23.30 and 07.00 (11.30 p.m. and 7 a.m.) in a built-up area. When your vehicle is stopped on the road you may only use your horn at times of danger due to another vehicle moving.

WAITING AND PARKING

115 You MUST NOT let your vehicle stand:

- on a motorway, except on the hard shoulder in an emergency;
- on the carriageway of any pedestrian crossing or within the area marked by zigzag lines on either side of a Zebra crossing or in the zone indicated by rows of studs on the approach to Pelican crossings — except to allow a pedestrian to cross;
- on the carriageway of a "Clearway" except in an emergency (see page 50);
- on the right-hand side of the carriageway at night except in a one-way street;
- on the carriageway or the verges of an "Urban Clearway" during the times shown on the signs (see page 50), other than for no longer than is necessary to let passengers board or alight.
- in a bus lane during the prohibited time displayed, except to load or unload goods where permitted.
- on the carriageway or the verges of any section of road marked with double white lines even if one of the lines is broken (see page 55), except to let passengers board or alight or to load or unload goods;
- on the side of the carriageway or on the pavement or verge along that side of a road where there are yellow lines near the edge (see page 56) during the times shown on the plates on that side of the road, except while passengers board or alight or while loading or unloading goods. In controlled parking zones the standard times applying throughout the zone are shown on the entry signs (see page 56) and roadside time plates are erected only where different times apply. Also, you must not wait to load or unload goods where there are marks on the kerb (see page 56) during the times shown on the "No

loading'' plates; the absence of a yellow line does not automatically mean that parking is permitted there. Certain concessions are allowed to disabled people, whose vehicles can be recognised by the orange badge (see page 60);

- in parking areas in a ''Disc Zone'' (indicated on signs) unless it displays a parking disc; the traffic signs at the entry roads to the zone and time plates at the parking places within the zone will indicate the days and times when the scheme is in operation, the time allowed for parking and the period during which return to that parking place is prohibited. These details may vary in different zones;
- in parking places reserved for specific users, for example residents or disabled badge holders — unless you are entitled to use these spaces;
- between 07.00 and 19.00 (7 a.m. and 7 p.m.) at bus stops marked by a wide yellow line unless it is a bus or coach authorised to stop there;
- on any verge, central reservation or footway, if it is a goods vehicle with a maximum laden weight (including any trailer) exceeding 7·5 tonnes, except in certain circumstances, for example, if loading or unloading could only be performed there and the vehicle is not left unattended.

116 Also, do not let your vehicle stand:

- where it would cause danger to other vehicles or pedestrians, for example—
at or near a school entrance or a school crossing patrol — not even to pick up or set down passengers;
where it would hide a traffic sign;
on a footpath, pavement or cycle path;
at or near any bus stop;on or near a level crossing;
- where it would make it difficult for others to see clearly, for example, near or at—
a junction, i.e. not within 15 yards of it;
a bend;
the brow of a hill;
a hump-back bridge;
- where it would make the road narrow, for example—
opposite a traffic island;
alongside another stationary vehicle;
opposite or nearly opposite another stationary vehicle if

this would narrow the road to less than the width of two vehicles;
near roadworks;

- where it would hold up traffic or inconvenience others, for example—
on a narrow road;
on flyovers, in tunnels or in underpasses;
on fast main roads, except in a lay-by;
on a single-track road, or in a passing place on such a road;
blocking a vehicle entrance to properties;
blocking the entrance to or exit from a car park;
where it would prevent the use of properly parked vehicles;

- where emergency vehicles stop or go in and out, for example—
hospital and ambulance entrances;
doctors' entrances;
police and fire stations;
fire hydrants;
entrances to coastguard stations.

Make sure you always park your vehicle safely and where it will cause the least inconvenience to others. Walk a few yards rather than cause accidents.

117 If your vehicle is fitted with a hazard warning device (i.e., a switch to permit all the direction indicators to flash simultaneously), it may only be used when the vehicle is stationary, to indicate that the vehicle is causing a temporary obstruction to traffic flow (for example, because it has broken down or is being loaded or unloaded). The device must not be used whilst the vehicle is in motion, nor should it be regarded as providing an excuse for stopping when you should not.

118 Always pull off the road on to a parking area if you can.

119 Before opening any door of a vehicle make sure that there is no one on the road, pavement or footpath close enough to be hit by the door. Be particularly careful about cyclists. Get out on the side nearer the kerb whenever you can and make sure that your passengers (especially children) do so too.

120 If you have to park on the road, stop as close as you can to the edge. Before leaving your vehicle make sure the handbrake is on firmly, switch off the engine and headlamps. Always lock your vehicle.

121 Never park on the road at night if it can be avoided. Except as described in Rule 122, it is illegal to park at night without lights and it is particularly dangerous to park on the road in fog. Lights should always be left on in these conditions.

122 Cars, goods vehicles not exceeding 30 cwt unladen, invalid carriages and motorcycles may be parked at night without lights on a road subject to a speed limit of 30 mph or less, but only if:

a the vehicle is at least 15 yards away from a junction, close and parallel to the kerb and facing in the direction of traffic flow; or—

b it is in a recognised parking place.

Trailers and vehicles with projecting loads must not be left without lights on a road at night.

BREAKDOWNS AND ACCIDENTS

123 If you have a breakdown, think first of other traffic. Get your vehicle off the road if possible and keep your passengers and yourself off the road.

124 Take steps to warn other drivers of an obstruction. If your vehicle is fitted with hazard warning lights, use them. If you carry a red warning sign (a reflecting triangle), place it on the road at least 50 yards (150 yards on the hard shoulder of motorways) before the obstruc-

tion and on the same side of the road. At night or in poor visibility do not stand at the rear of your vehicle or allow anyone else to do so — you may obscure the rear lamps.

125 If anything falls from your vehicle, stop as soon as you can with safety and remove it from the carriageway.

126 There may have been an accident if you see several vehicles in the distance which are going very slowly or have stopped, or if you see warning signs and the flashing lights of emergency vehicles. Slow down and be prepared to stop.

127 If you are first on the scene of an accident you should:

a warn other traffic by displaying a red triangle or switching on hazard warning lights or other lights or by any other means. Extinguish lighted cigarettes or other fire hazards and ask drivers to switch off their engines;

b arrange for the police and ambulance authorities to be summoned immediately with full details of the location and casualties; on a motorway, if necessary drive on to the next emergency telephone;

c remove casualties if in any further immediate danger but do not move them unnecessarily; give first aid as described on page 70;

d get uninjured people out of the vehicles and into a place of safety; on a motorway this should be away from the carriageway or hard shoulder or central reservation;

e stay at the scene until emergency services arrive.

Accidents involving dangerous goods

128 If the accident involves a vehicle containing dangerous goods (the vehicle may display a hazard information panel as shown on page 59 and often will have other information such as the name of the substance being carried), you should also:

a arrange for the police or fire brigade to be given immediately as much additional information as possible about the labels and other markings;

b keep everyone well away from the vehicle; even if you act to save life do so with the utmost caution as dangerous liquids may be leaking on to the highway; beware also of dangerous dust or vapours being carried towards you by the wind.

Extra rules for cyclists

129 Make sure your cycle is safe to ride. At night you must have front and rear lamps and a rear reflector. Your brakes, lamps and reflector must be kept in proper working order. Make sure your tyres are in good condition and are properly pumped up. It is a good idea to fit a bell to your cycle and to use it, if necessary, to warn other people on the road that you are coming.

130 Before starting to ride, always look round and make sure that it is safe to move away from the kerb. Before turning right or left, moving out to pass or pulling up at the kerb, always glance behind and make sure it is safe. Give a clear arm signal to show what you intend to do.

131 Do not ride more than two side by side. Ride in single file on busy narrow roads. You must not ride on a footpath or pavement by the side of the road.

132 On busy roads and at night, if you want to turn right it is often safer to stop first on the left-hand side of the road. Wait for a safe gap in the traffic before you start to turn.

133 You may only use a bus lane where the signs show the symbol of a bicycle. You must not use other bus lanes.

134 You may, if you wish, follow the procedure in Rules 102—105 for roundabouts. But if, because of inexperience or for any other reason, you feel unable to do so, you should either stay in the left-hand lane of the roundabout and look out particularly for vehicles crossing your path to leave the roundabout, or get off your cycle and walk.

135 Remember that you cannot be seen as easily as larger vehicles and that you should always give clear arm signals to let drivers behind you know what you intend to do, especially at roundabouts and junctions.

136 When you are riding:

a always keep both hands on the handlebar unless you are signalling;

b always keep both feet on the pedals;

c do not hold on to another vehicle or another cyclist;

d do not carry a passenger unless your cycle has been built or altered to carry one;

e do not ride close behind another vehicle;

f do not carry anything which might affect your balance or become entangled with the wheels or chain;

g do not lead an animal;

h wear light-coloured or reflective and fluorescent clothing (see Rule 4).

137 If there is a suitable cycle path, ride on it.

The road user in charge of animals

138 Do not let your dog out on its own. When you take it for a walk on the road, keep it on a lead.

139 If you have an animal in your car, keep it under control. Make sure it cannot disturb you while you are driving. Do not let a dog out of a car on to the road unless it is on a lead.

140 Before riding a horse on the roads, make sure you can control it in traffic. When riding, keep to the left. If you are leading a horse, on foot or while riding another, you should also keep to the left and keep the led animal on your left. In one-way streets, proceed only in the direction of the traffic and keep to the left. You must not ride, lead or drive a horse on a footpath or pavement by the side of the road.

141 If you are riding a horse, you should wear a hard hat.

142 If you are riding or herding animals after sunset, you should wear light-coloured or reflective clothing and carry lights which show white to the front and red to the rear.

143 If you are herding animals, keep to the left of the road and if there is someone with you send him along the road to warn drivers at places such as bends and brows of hills where they may not be able to see. If your herd is very large, divide it into smaller groups.

144 Take particular care at level crossings (see Rules 176–185).

Motorway driving

Rules 26—36, 37 (first part), 40, 44—48, 50, 69 (part), 70, 72, 78—80, 83, 111, 111[A], 112 (part), 113—114, 115 (part), 123—124, 126—128 and 139 also apply to motorway driving.

GENERAL

145 Motorways are dual-carriageway roads which must not be used by pedestrians, learner drivers, cyclists and riders of small motorcycles. Slow-moving vehicles, agricultural vehicles and some carriages used by invalids are also prohibited. (See page 69.) It is an offence to pick up or set down a passenger or a hitch-hiker on any part of a motorway including a slip road.

146 Traffic travels faster on motorways than on ordinary roads and you will need to sum up traffic situations more quickly. Using your mirror and concentrating all the time are doubly important on motorways.

147 Make sure that your vehicle is fit to cruise at speed, has correct tyre pressures for motorway driving and enough petrol, oil and water to take you at least to the next service area. You must ensure that any loads carried or towed are secure.

148 Slip roads and link roads between motorways may have sharp bends which you can take safely only by reducing speed.

JOINING THE MOTORWAY

149 When you join the motorway other than at its start, you will normally approach from a road on the left (a slip road). Give way to traffic already on the motorway. Watch for a safe gap in the traffic in the left-hand lane on the motorway and then adjust your speed in the extra lane (the acceleration lane) so that when you join it you are already travelling at the same speed. If there is not a suitable gap, wait in the acceleration lane until it is safe to enter the motorway.

150 After joining the motorway, stay in the left-hand lane long enough to get used to the speed of traffic before trying to overtake.

ON THE MOTORWAY

151 You must not reverse or turn in the road, or cross the central reservation, or drive against the traffic.

152 Even if you have missed your turn-off point or have taken the wrong route you must carry on until you reach the next exit.

153 In good visibility and weather conditions, drive at a steady cruising speed within the limits of your vehicle. You must not break the speed limits for the motorways or for your vehicle. On wet or icy roads, or in fog, keep your speed down.

154 Driving for long distances may make you feel sleepy. To help prevent this, make sure there is plenty of fresh air in your vehicle or stop at a service area, or turn off at an exit, and walk around.

Lane discipline

155 On a two-lane carriageway, drive in the left-hand lane except when overtaking.

156 On carriageways with three or more lanes the normal ''Keep to the left'' rule still applies. You may, however, stay in the middle lane when there are slower vehicles in the left-hand lane, but you should return to the left-hand lane when you have passed them. The right-hand lane is for overtaking only. If you use it, move back to the middle lane and then into the left-hand lane as soon as you can, but without cutting in.

157 To help drivers on motorways, there are amber-coloured studs marking the right-hand edge of the carriageway, red studs marking the left-hand edge and green studs separating the acceleration and deceleration lanes from the through carriageway.

158 On some motorways, direction signs may be placed over the road. Pay special attention to the signs and move into the correct lane in good time.

159 A goods vehicle which has an operating weight of more than 7·5 tonnes, or any vehicle drawing a trailer, or a bus longer than 12 metres, must not use the right-hand lane of a carriageway with three or more lanes unless there are exceptional circumstances.

Motorway fog

160 When driving in fog, it is vital that you should obey the rules in Rule 50.

Overtaking

161 Overtake only on the right, unless traffic is moving in

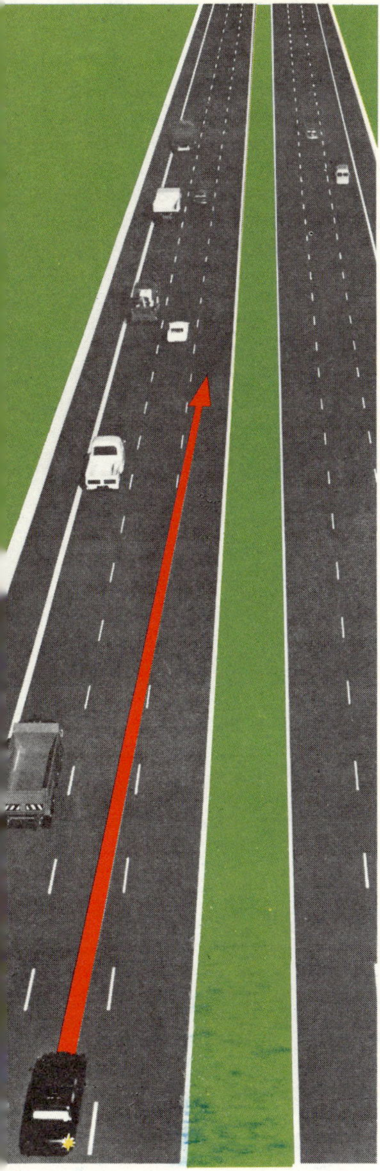

queues and the traffic queue on your right is moving more slowly than you are. Never move to a lane on your left to overtake. Never use the hard shoulder for overtaking.

162 Do not overtake unless you are sure it is safe for yourself and others. Many accidents on motorways are rear-end collisions. So before you start to overtake make sure that the lane you will be joining is clear far enough behind — use your mirror — and ahead. Remember that traffic may be coming up behind much more quickly than you think. Signal before you move out. Be particularly careful at dusk, in the dark and in fog or mist, when it is more difficult to judge speed and distance.

Remember: **Mirror — Signal — Manoeuvre.**

163 Get back to the left-hand lane or, if this is occupied, the middle lane as soon as you can after overtaking, but do not cut in on the vehicle you have just overtaken.

Breakdowns

164 If you have a breakdown, get your vehicle off the carriageway on to the hard shoulder so that it does not get in the way of other traffic. Never forget the danger from passing traffic, so do not step on to the carriageway or let your passengers do so. Use the roadside telephone to inform the police. Do not delay in obtaining assistance and do not leave your vehicle unattended for a long period. Signs on the marker posts indicate the direction of the nearest telephone. If you cannot move your vehicle off the carriageway, take steps to warn other drivers of its presence (see Rule 124). When rejoining the carriageway, build your speed up first on the hard shoulder. Watch for a safe gap in the traffic before rejoining it.

Obstructions

165 If anything that may cause danger falls from your vehicle or from another vehicle, use the roadside telephone to inform the police. Do not try to retrieve it yourself.

Motorway signals

166 Special signals are used on motorways. In normal conditions they are blank. In dangerous conditions, amber lights flash and a panel in the middle of the signal shows either a special temporary maximum speed or which lanes are closed. When the danger has been passed the panel of the next signal will show (without flashing lights) the end of restriction signal.

167 On most motorways, the signals are on the central reservation at intervals of not more than two miles and they apply to all lanes. On some very busy motorways, the signals are overhead, one applying to each lane.

168 Some signals have red lights as well. If the red lights above your lane flash, you must not go beyond the signal in that lane. If red lights flash on a slip road, you must not enter it.

169 Some motorways still have flashing amber signals at their entrances and at one- or two-mile intervals. These warn of danger; for example, an accident, fog or risk of skidding. When the signals are flashing, keep your speed under 30 mph until you are sure it is safe to go faster.

170 These signals are for your safety. Always act on them. Remember — danger may be present even if you cannot see the cause.

ROADSIDE SIGNALS

Temporary maximum speed

Lane closed ahead

End of restriction

OVERHEAD SIGNALS

1 

2 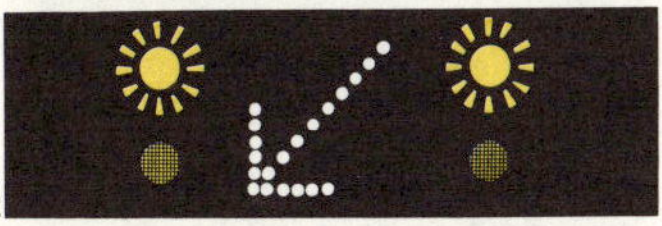

3 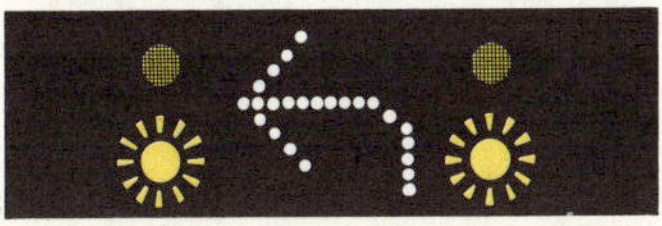

4 

5 

1 Temporary maximum speed
2 Change lane
3 Leave motorway at next exit
4 Do not proceed any further in this lane
5 End of restriction

Stopping and parking

171 You MUST NOT stop except:

a in an emergency (for example, to prevent an accident);

b when you break down (see Rules 123 and 164);

c when you are signalled to do so by the police, by an emergency traffic sign or by a flashing red light signal.

172 You may park only at a service area. You must not park on:

a the carriageway itself;

b the slip roads;

c the hard shoulders (except in an emergency);

d the central reservation.

173 You must not walk on the carriageway. In an emergency be particularly careful to keep children and animals off the carriageway and the hard shoulders.

LEAVING THE MOTORWAY

174 If you are not going to the end of the motorway you will normally leave by a slip road on your left. Watch for the signs letting you know you are getting near your turn-off point. If you are not already in the left-hand lane, move into it well before reaching your turn-off point and stay in it. Give a left turn signal in good time and, if necessary, reduce speed. Then get into the extra lane provided (the deceleration lane) where you can slow down before you join the slip road.

175 When leaving the motorway or using a link road remember to adjust your driving to suit the new conditions. Your speed will be higher than you think — 50 mph may feel like 30 mph — so be sure to use your speedometer.

The road user and railway level crossings

GENERAL

176 Approach a level crossing at a moderate speed and cross it with care. Do not loiter. Never drive "nose to tail" over it. Never drive on to one unless you can see the road is clear on the other side. NEVER STOP ON OR IMMEDIATELY BEYOND ANY LEVEL CROSSING.

LEVEL CROSSINGS WITH GATES OR FULL BARRIERS

177 Many level crossings have gates, or barriers with skirts, that are operated either by an attendant or by remote control and go right across the road. Some also have amber lights followed by flashing red "STOP" lights. Do not pass the lights once they show. If there are no lights at all, stop when the gates begin to close or when the barriers start to descend.

178 Some level crossings with gates or barriers but no attendant have "STOP" signs and small red and green lights. Do not cross when the red light is showing, as a train is coming. There is a special railway telephone and if you are driving a very large or slow-moving vehicle, or are herding animals, you must first telephone the signalman to make sure it is safe for you to cross. Open both gates or fully raise both barriers before you cross and close or lower them after you. When you have crossed, telephone the signalman to let him know you are over.

179 Some level crossings have gates, but no attendant or red lights. At such crossings, stop, look both ways, listen and make sure there is no train coming. If there is a special railway telephone, first telephone the signalman to make sure it is safe for you to cross. Before you cross with a vehicle or with animals, open *both* gates wide and then make a further check that no train is coming. Drive your vehicle or animals clear of the crossing and then close both gates.

OPEN LEVEL CROSSINGS

180 At level crossings with no gates, barriers, attendant or red lights, there will be a "Give Way" or "STOP" sign. You must look both ways, listen and make sure there is no train coming before you cross. Always "Give Way" to trains.

AUTOMATIC OPEN CROSSINGS

181 Some level crossings with no gates, barriers or attendant have amber lights followed by flashing red "STOP" lights. When the lights show you must stop and wait. Do not cross the railway – a train will reach the crossing soon after the lights begin to show. If one train has gone by, but the lights continue to flash, you must wait as another train will soon arrive. The lights will go out when it is safe to cross.

AUTOMATIC HALF-BARRIER LEVEL CROSSINGS

182 These crossings have automatic barriers across the left side of the road. The trains work the barriers, which descend just before the train reaches the crossing. Amber lights and an alarm followed by flashing red "STOP" lights warn you when the barriers are about to come down. Do not move on to the railway once these signals have started – the train cannot stop and will be at the crossing very soon. Wait at the "STOP" line. If you are on foot, wait at the barrier, or the broken white line on the road or footpath, or by the wicket gate. Never zigzag round the barriers – you could be killed and endanger other lives. If one train has gone by, but the barriers stay down and the red lights continue to flash, you must wait as another train will soon arrive.

183 If you are already crossing when the amber lights and alarm start, KEEP GOING.

184 If the barriers stay down at any time for more than three minutes without a train arriving, use the telephone at the crossing to ask the signalman's advice. If you are driving a large or slow-moving vehicle you must first telephone the signalman, as instructed by the sign before the crossing, to get his permission to cross. There is a special railway telephone at the sign. If you are herding animals, you must also telephone for permission to cross. If you have telephoned the signalman before crossing, telephone him again to tell him you are over.

185 If your vehicle stalls, or breaks down, or if you have an accident on the crossing:

First: get your passengers out of the vehicle and clear of the crossing; then use the telephone at the crossing immediately to tell the signalman.

Second: if there is time, move the vehicle clear of the crossing.

If the amber lights and alarm start, stand well clear of the crossing.

SIGNS AT LEVEL CROSSINGS

Before the crossing

At an open level crossing

Warning of low ground clearance at crossing

Light signals controlling traffic

TRAFFIC LIGHT SIGNALS

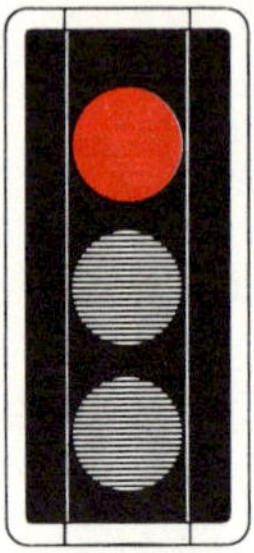

RED means "Stop". Wait behind the stop line on the carriageway.

RED AND AMBER also means "Stop". Do not pass through or start until GREEN shows.

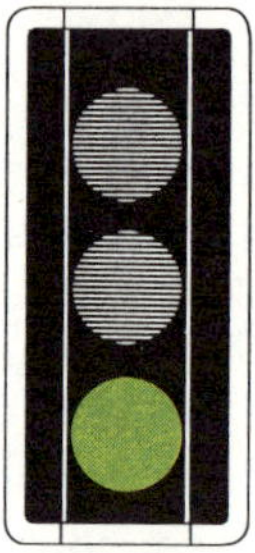

GREEN means you may go on if the way is clear. Take special care if you mean to turn left or right and give way to pedestrians who are crossing.

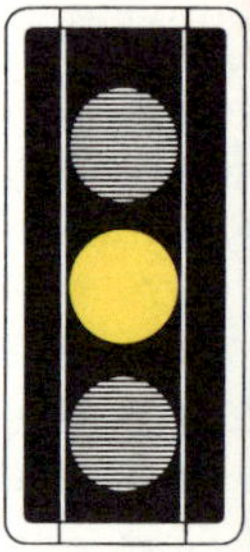

AMBER means "Stop" at the stop line. You may go on only if the AMBER appears after you have crossed the stop line or are so close to it that to pull up might cause an accident.

A GREEN ARROW may be provided in addition to the full green signal if movement in a certain direction is allowed before or after the full green phase. If the way is clear you may go but only in the direction shown by the arrow. You may do this whatever other lights may be showing.

FLASHING RED LIGHTS

Alternately flashing red lights mean YOU MUST STOP

At level crossings, lifting bridges, airfields, fire stations, etc

Signals by authorised persons

Stop

Vehicle approaching from behind

Vehicle approaching from the front

Vehicles approaching from both front and behind

Come on

Beckoning on a vehicle from the front

Beckoning on a vehicle from the side

Beckoning on a vehicle from behind

Signals to other road users

DIRECTION INDICATOR SIGNALS

I intend to
move out to
the right
or turn right

I intend to
move in to
the left
or turn left
or stop on
the left

STOP LIGHT SIGNALS

I am slowing
down or
stopping

These signals should not be used except for the purposes described.

ARM SIGNALS

For use when direction indicator signals are not used; or when necessary to reinforce direction indicator signals and stop lights. Also for use by pedal cyclists and those in charge of horses.

I intend to move out to the right or turn right

I intend to move in to the left or turn left

I intend to slow down or stop

This signal is particularly important at Zebra crossings to let other road users, including pedestrians, know that you are slowing down or stopping.

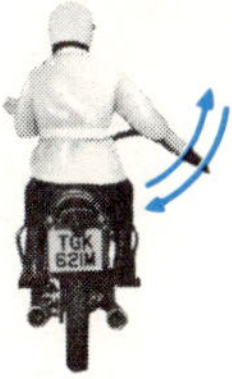

ARM SIGNALS TO PERSONS CONTROLLING TRAFFIC

I want to go straight on

I want to turn left

I want to turn right

Traffic Signs

SIGNS GIVING ORDERS

These signs are mostly circular and those with red circles are mostly prohibitive

Maximum speed

National speed limit applies

Stop and Give Way

Give way to traffic on major road

School crossing patrol

No vehicles

No entry for vehicular traffic

No right turn

No left turn

No U turns

No overtaking

Give priority to vehicles from opposite direction

No motor vehicles

No motor vehicles except solo motorcycles, scooters or mopeds

Manually operated temporary 'STOP' sign

No vehicles with over 12 seats •except regular scheduled school and works buses

No cycling

No pedestrians

No goods vehicles over maximum gross weight shown (in tonnes)

No vehicles including load over weight shown (in tonnes)

Axle weight limit in tonnes

No vehicles over height shown

No vehicle or combination of vehicles over length shown

No vehicles over width shown

No stopping (Clearway)

Parking restricted to use by people named on sign

No stopping during times shown except for as long as necessary to set down or pick up passengers

Plates below some signs qualify their message

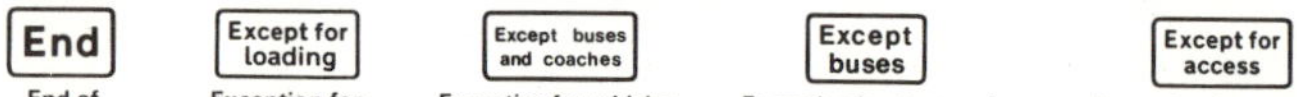

End of restriction

Exception for loading/unloading goods

Exception for vehicles with over 12 seats

Exception for stage and scheduled express carriages, school and works buses

Exception for access to premises and land adjacent to the road where there is no alternative route

---

Signs with blue circles but no red border mostly give positive instruction

Ahead only

Turn left ahead (right if symbol reversed)

Turn left (right if symbol reversed)

Keep left (right if symbol reversed)

Vehicles may pass either side to reach same destination

Route to be used by pedal cyclists only

Minimum speed

End of minimum speed

Mini-roundabout (roundabout circulation – give way to vehicles from the immediate right)

One-way traffic (Note: compare circular "Ahead only" sign)

Shared pedal cycle and pedestrian route

With-flow bus and cycle lane

Contra-flow bus lane

With-flow pedal cycle lane

WARNING SIGNS *Mostly triangular*

Distance to "STOP" line ahead

Cross roads

Roundabout

T junction

Staggered junction

Distance to "Give Way" line ahead

Double bend first to left (may be reversed)

Plate below some signs

Sharp deviation of route to left (or right if chevrons reversed)

Bend to right (or left if symbol reversed)

Dual carriageway ends

Slippery road

Two-way traffic straight ahead

Two-way traffic crosses one-way road

Traffic merges from left/right with equal priority

Road narrows on right (left if symbol reversed)

Road narrows on both sides

Crossing point for elderly people (blind or disabled if shown)

Pedestrians in road ahead

Pedestrian crossing

Children going to or from school

School crossing patrol ahead (Some signs have amber lights which flash when patrol is operating)

Uneven road

Traffic signals

Failure of light signals

Steep hill downwards

Steep hill upwards

Gradients may be shown as a ratio i.e. 20% = 1:5

Risk of grounding of long low vehicles at level crossing

Road works

Hump bridge

Change to opposite carriageway (may be reversed)

Loose chippings

Worded warning sign

Plate to indicate a level crossing equipped with automatic barriers and flashing lights

Level crossing with barrier or gate ahead

Level crossing without barrier or gate ahead

Level crossing without barrier (the additional lower half of the cross is used when there is more than one railway line)

Cycle route ahead

Height limit (e.g. low bridge)

Available width of headroom indicated

Opening or swing bridge ahead

Quayside or river bank

Overhead electric cable; plate indicates maximum height of vehicles which can pass safely

Cattle

Wild animals

Wild horses or ponies

Accompanied horses or ponies crossing the road ahead

Other danger; plate indicates nature of danger

Distance to tunnel

Falling or fallen rocks

Low-flying aircraft or sudden aircraft noise

Distance over which road humps extend

DIRECTION SIGNS *Mostly rectangular*

Signs on motorways *Blue backgrounds*

Start of motorway and point from which motorway regulations apply

On approaches to junctions
(junction number on black background)

Route confirmatory sign after junction

End of motorway

At a junction leading directly into a motorway

Downward pointing arrows mean "Get in lane"

The panel with the sloping arrow indicates the destinations which can be reached by leaving the motorway at the next junction

On approaches to junctions

Ring road

Signs on primary routes

Green backgrounds

At the junction

Route confirmatory sign after junction

On approaches to junctions (The blue panel indicates that the motorway commences from the junction ahead. The motorway shown in brackets can also be reached by proceeding in that direction)

Route confirmatory sign after junction

Signs on non-primary routes

Black borders

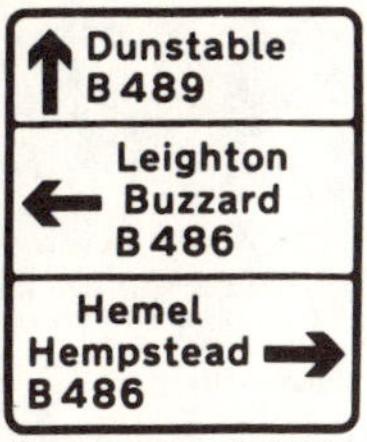

On approaches to junctions

Ring road

At the junction

Local direction signs

Blue borders

On approaches to junctions (where there is a different route for pedal cycles this may be shown in a blue panel)

On approaches to junctions

At the junction

Direction to camping and caravan site

Tourist attraction

Holiday route

Diversion route

English heritage

Direction to toilets with access for the disabled

Advisory route for lorries

Route for pedestrians

Picnic site

Recommended route for pedal cycles to place shown

Airport

INFORMATION SIGNS *All rectangular*

Entrance to controlled parking zone

One-way street

Parking place for towed caravans

Priority over vehicles from opposite direction

No through road

Hospital ahead

Advance warning of restriction or prohibition ahead

Distance to service area with fuel, parking and cafeteria facilities (The current petrol price may be shown in pence per gallon or litre, or may be omitted)

End of controlled parking zone

Appropriate traffic lanes at junction ahead

"Count-down" markers at exit from motorway (each bar represents 100 yards to the exit) Green-backed markers may be used on primary routes

Recommended route for pedal cycles

Tourist information point

The number of arrows and red bars may be varied according to the number of lanes open and closed

Bus lane on road at junction ahead

Lane control signals

White arrow – lane available to traffic facing the sign. Red crosses – lane closed to traffic facing the sign

Road markings

ACROSS THE CARRIAGEWAY

Give way to traffic on major road

Give way to traffic from the right in roundabout

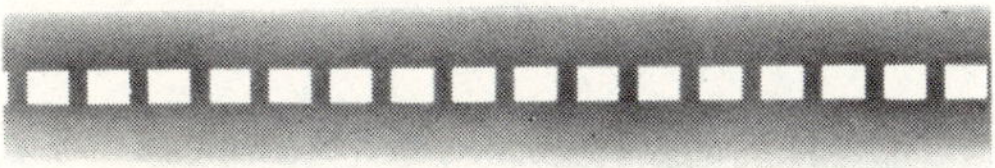

Give way to traffic from right at mini-roundabout

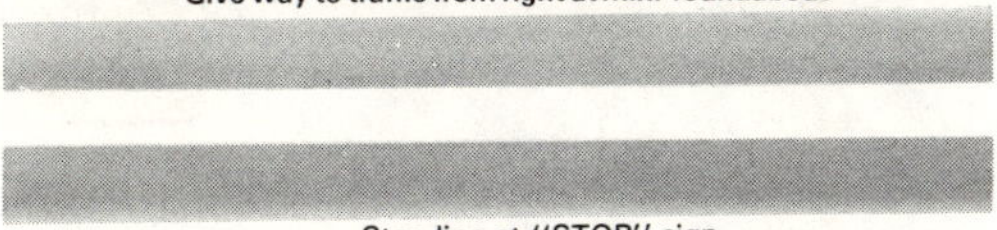

Stop line at ''STOP'' sign

Stop line at signals or police control

ALONG THE CARRIAGEWAY

Double white lines

See Rules 66 and 67

Diagonal stripes

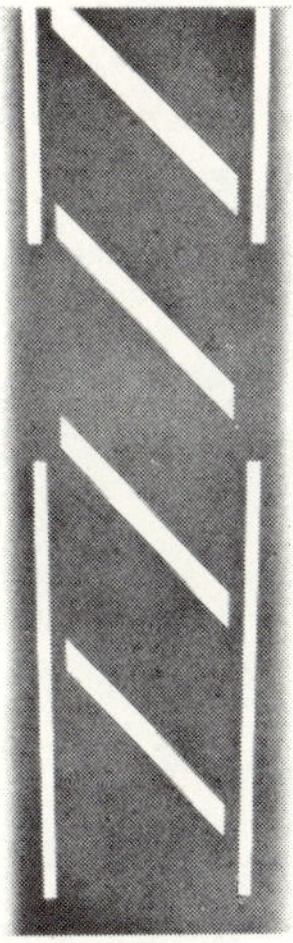

See Rule 68

Lane markings

Lane line
See Rules
69 and 70

Centre line

Hazard
warning line
See Rule 65

ALONG THE EDGE OF THE CARRIAGEWAY

Waiting restrictions

No waiting on carriageway, pavement or verge (except to load or unload or while passengers board or alight) at times shown on nearby plates or on entry signs to controlled parking zones.

If no days are indicated on the sign, the restrictions are in force every day including Sundays and Bank Holidays. The lines give a guide to the restriction in force but the time plates must be consulted.

No waiting for at least eight hours between 7 am and 7 pm on four or more days of the week

No waiting for at least eight hours between 7 am and 7 pm on four or more days of the week plus some additional period outside these times

During any other periods

Examples of plates indicating restriction times

Continuous prohibition

Plate giving times

Limited waiting

ON THE KERB OR AT EDGE OF CARRIAGEWAY

Loading restrictions

No loading or unloading at times shown on nearby plates. If no days are indicated on the sign the restrictions are in force every day including Sundays and Bank Holidays.

During every working day

For example

During every working day, and additional times

For example

During any other periods

For example

ZEBRA CONTROLLED AREAS

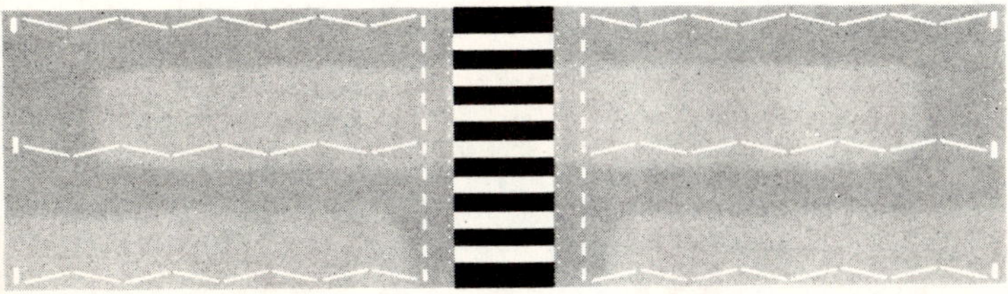

OTHER ROAD MARKINGS

Keep entrance clear of stationary vehicles, even if picking up or setting down children

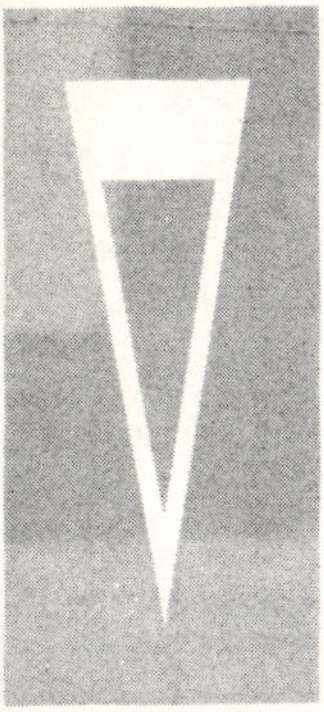

Warning of "Give Way" just ahead

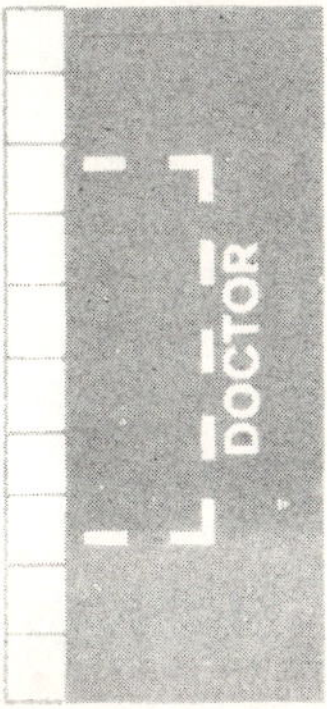

Parking space reserved for vehicles named

See Rule 115

See Rule 77

Box junction
(see Rule 92)

Do not block entrance to side road

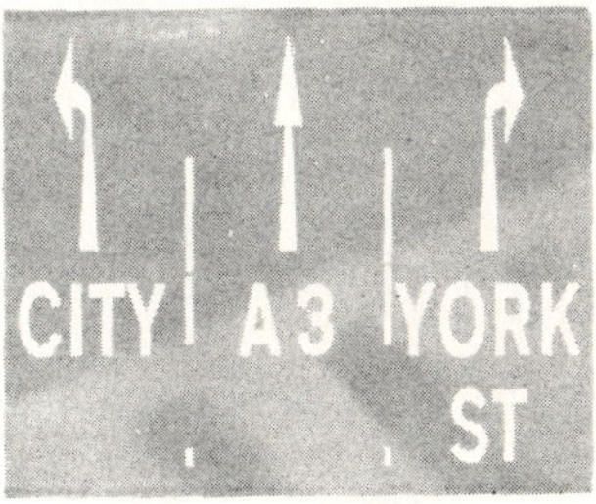

Indication of traffic lanes

Note: Although *The Highway Code* shows many of the signs commonly in use, a comprehensive explanation of our signing system is given in the Department's booklet *Know Your Traffic Signs*, which is on sale at booksellers. The booklet also illustrates and explains the vast majority of signs the road user is likely to encounter.
The signs illustrated in *The Highway Code* are not all drawn to the same scale. In Wales, bilingual versions of some signs are used.

Vehicle markings

HEAVY GOODS VEHICLE REAR MARKINGS

Motor-vehicles over 7500 kilograms maximum gross weight and trailers over 3500 kilograms maximum gross weight

Left

Rigł

The vertical markings are also required to be fitted to builders' skips placed in the road. Commercial vehicle or combination longer than 13 metres (optional on combinations between 11 and 13 metres)

Left

Right

or

PROJECTION MARKERS

End marker

Side marker

Both required when load or equipment (e.g. crane jib) overhangs front or rear by more than 1.83 metres

TRANSPORT HAZARD INFORMATION SYSTEM

Certain vehicles carrying dangerous goods must display hazard information panels:

The panel illustrated is for a flammable liquid.
Diamond symbols indicating other risks include:

Toxic substance

Oxidizing substance

Corrosive substance

Spontaneously combustible substance

Radioactive substance

Non-flammable compressed gas

IF A SPILLAGE OCCURS –
KEEP WELL AWAY AND INFORM THE POLICE OR THE FIRE BRIGADE

ORANGE BADGE SCHEME

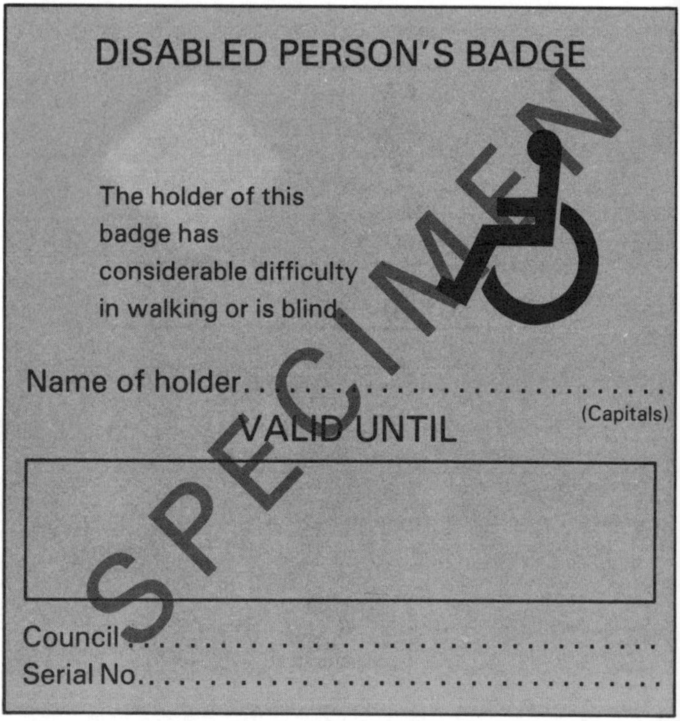

Windscreen badge for disabled persons entitled to parking concessions (see Rule 115)

The law's demands

The following pages deal with major points of the law affecting safety on the roads. For the precise wording of the law please refer to the various Acts and Regulations. These are indicated in the margin by the following abbreviations.

BP(S)A	Burgh Police (Scotland) Act 1892.*
CG(S)A	Civic Government (Scotland) Act 1982.
CUR	Motor Vehicles (Construction and Use) Regulations 1978.
DS(RTTC)R	Dangerous Substances (Conveyance by Road in Road Tankers and Tank Containers) Regulations 1981.
GV(PT)R	Goods Vehicles (Plating and Testing) Regulations 1971.
HA	Highway Act 1835, or, as the case may be, Highways Act 1980.
HCVA	Heavy Commercial Vehicles (Controls and Regulations) Act 1973.
HGV(DL)R	Heavy Goods Vehicles (Drivers' Licences) Regulations 1977.
LA	Licensing Act 1872.
MC(PH)R	Motorcycles (Protective Helmets) Regulations 1980.
MT(E&W)R	Motorways (England & Wales) Regulations 1982.
MT(E&W)(A)R	Motorways Traffic (England and Wales) (amendment) Regulations 1984.
MT(SL)R	Motorways Traffic (Speed Limit) Regulations 1974.
MT(S)R	Motorways Traffic (Scotland) Regulations 1964.
MV(T)R	Motor Vehicles (Tests) Regulations 1981.
MV(DL)R	Motor Vehicles (Driving Licences) Regulations 1981.
MV(WSB)R	Motor Vehicles (Wearing of Seat Belts) Regulations 1982.
MV(WSBC)R	Motor Vehicles (Wearing of Seat Belts by Children) Regulations 1982.
PCUR	Pedal Cycles (Construction and Use) Regulations 1983.
PPCRGD	"Pelican" Pedestrian Crossings Regulations and General Directions 1969.
PPVA	Public Passenger Vehicles Act 1981.
RB(S)A	Roads and Bridges (Scotland) Act 1878.
RTA	Road Traffic Acts 1972 and 1974.
RTRA	Road Traffic Regulation Act 1984.
RVLR	Road Vehicles Lighting Regulations 1984.
RV(R and L)R	The Road Vehicles (Registration and Licensing) Regulations 1971.
SRA	Special Roads Act 1949.
SRCTO	Special Roads (Classes of Traffic) Order 1971.
SRCT(S)O	Special Roads (Classes of Traffic) (Scotland) Order 1971.
TSRGD	Traffic Signs Regulations and General Directions 1981.
TA	Transport Acts 1980 and 1982.
VEA	Vehicles (Excise) Act 1971.
ZPCR	"Zebra" Pedestrian Crossings Regulations 1971.

** This Act does not apply within Edinburgh, Glasgow, Aberdeen, Dundee and Greenock, where local Acts make different provisions.*

To pedestrians

	Acts & Regulations
You have precedence when you are on the carriageway within the limits of an uncontrolled Zebra crossing, and on a Pelican crossing when the signal to cross is illuminated. *NOTES: (a) An uncontrolled Zebra crossing is one at which traffic is not being controlled by a police constable * and which is marked with two or more lighted beacons, black and white stripes, and studs to indicate the limits of the crossing.* *(b) You have NO precedence when you are standing on the kerb or when you are standing on a street refuge or central reservation which is on a Zebra crossing, or on a Pelican crossing when the RED MAN symbol is illuminated.*	**PPCRGD** No 7 and **ZPCR** No 8

You must not

— loiter on any type of pedestrian crossing:	**PPCRGD** No 12 and **ZPCR** No 9
— wilfully obstruct the free passage along a highway;	**HA 1980** Sec 137 **RB(S)A** Sch C and **BP(S)A** Sect 381 (10)
— walk on motorways or their slip roads;	**MT(E&W)R** No 13 **MT(S)R** No 11
— proceed along or across the carriageway when given a direction to stop by a police constable * engaged in controlling traffic;	**RTA 1972** Sect 23
— without lawful authority or reasonable cause, hold on to or get on a motor vehicle or trailer in motion or tamper with the brake or other part of the mechanism of a motor vehicle;	**RTA 1972** Sects 29 and 30
— be drunk in any highway or public place.	**LA** Sect 12 and **CG(S)A** Sect 50

To dog owners

You must not

— allow your dog to be off its lead on a road which has been designated as one where dogs must be kept on a lead, unless your dog is kept for tending sheep or cattle or is in use under proper control for sporting purposes.	**RTA 1972** Sect 31

To horse riders

You must not

— wilfully ride, lead or drive your horse on a footpath by the side of any road made or set apart for the use of foot passengers.	**HA 1835** Sect 72 **RB(S)A 1878** Sch C and **BP(S)A** Sec 381 (13)

To pedal cyclists

Before cycling, make sure that

— your cycle has efficient brakes.	**PCUR** Nos 7—9

**The reference to a police constable includes a traffic warden (Transport Act 1968).*

You must, even if you are wheeling your cycle,

— observe amber* and red "STOP" signals, traffic signs which give orders, double white lines (solid or broken), yellow road markings and the directions of a police constable † controlling traffic;	**RTA 1972** Sect 22 **PPCRGD** No 10 **RTRA** Sects 1, 6, 9 and 53
— stop when signalled to do so by a School Crossing Patrol exhibiting a "STOP — CHILDREN" sign;	**RTRA** Sect 28
— give precedence to pedestrians on an uncontrolled Zebra crossing, that is, a crossing marked by black and white stripes, studs and lighted beacons and at which there is no police constable † controlling traffic;	**ZPCR** No 8
— give precedence to pedestrians on a Pelican crossing, when an amber light is flashing.	**PPCRGD** No 11

You must

— at night, see that your front and rear lamps are alight and that your cycle has an efficient red rear reflector;	**RVLR** Nos 16, 20 and 21
— at night, if you are wheeling your cycle or are stationary without lights, keep as close as possible to the nearside edge of the road.	**RVLR No** 21
— stop when required to do so by a police constable † in uniform.	**RTA 1972** Sect 159

You must not

— stop your cycle within the limits of a pedestrian crossing, except in circumstances beyond your control or when it is necessary to do so to avoid an accident;	**PPCRGD** No 12 and **ZPCR** No 9
— on the approach to an uncontrolled Zebra crossing marked by a pattern of zigzag lines, overtake the moving motor vehicle nearest to the crossing or the leading vehicle which has stopped to give way to a pedestrian on the crossing,	**ZPCR** No 10
— ride recklessly;	**RTA 1972** Sect 17
— ride without due care and attention or without reasonable consideration for other persons using the road;	**RTA 1972** Sect 18
— ride under the influence of drink or a drug;	**RTA 1972** Sect 19
— wilfully ride on a footpath by the side of any road made or set apart for the use of foot passengers;	**HA 1835** Sect 72 **RB(S)A** Sch C and **BP(S)A** Sect 381 (40)
— by negligence or misbehaviour interrupt the free passage of any road user or vehicle;	**HA 1835** Sect 78 **BP(S)A** Sect 381 (10)
— leave your cycle on any road in such a way that it is likely to cause danger to other road users;	**RTA 1972** Sect 24
— leave your cycle where waiting is prohibited;	**RTRA** Sects 1, 6 and 9
— carry a passenger on a bicycle not constructed or adapted to carry more than one person;	**RTA 1972** Sect 21
— hold on to a motor vehicle or trailer in motion on any road.	**RTA 1972** Sect 30

See the illustrations on page 46.

† *The reference to a police constable includes a traffic warden (Transport Act 1968).*

To drivers of motor vehicles

Before driving, make sure that

	Acts & Regulations
— your vehicle is properly licensed and the tax disc displayed; (this is necessary even if it is only parked on a public road);	**VEA** Sects 1 and 12 (4)
— your use of the vehicle is properly insured and that there are no restrictions in the relevant insurance policy (for example, as to who may drive it) which would make your use of the vehicle illegal;	**RTA 1972** Sect 14
— you have a driving licence valid for the class of vehicle which you intend to drive, that it is not out of date, and that you have signed it in ink;	**RTA 1972** Sect 84 **MV(DL)R** No 11 an **HGV(DL)R** No 7 **PPVA** Sect 22,
— a provisional licence will not show entitlement to ride a motor cycle unless you request it;	**TA 1980** Sect 33, **RTA 1972** Sect 88.1(a), 96, 112 **MV(T)R**
— you have a current test certificate for your vehicle if it is over the prescribed age limit, and, where applicable, you have a current plating certificate for your goods vehicle;	**RTA 1972** Sects 43, 44 and 47 **GV(PT)R**
— your eyesight is up to the standard required for the driving test;	**RTA 1972** Sect 91 **MV(DL)R 1981** No 22 (1) (f)
— you have reported to the licensing authority any health condition likely to affect your driving;	**RTA 1972** Sect 87
— the condition of your vehicle, of any trailer it is drawing and of any load, and the number of passengers and the way in which they are carried, are such that they do not endanger yourself or others;	**CUR** NO 97 (1)
— your brakes and steering are in good working order and properly adjusted;	**CUR** Nos 101, 101A and 102
— your tyres are suitable for the vehicle, are properly inflated, have a tread depth of at least 1 mm and are free from cuts and other defects;	**CUR** Nos 107 and 108
— your windscreens and other windows comply with regulations concerning visual transmission of light and freedom from obstruction to vision; and are kept clean, where appropriate, in conjunction with windscreen wipers and washers which should be maintained in effective working order at all times;	**CUR** Nos 27, 28, 100 and 103
— your seat belts, anchorages, fastenings and adjusting devices are maintained free from obvious defects;	**CUR** No 102A
— your vehicle is fitted with the appropriate number of mirrors, so fitted that you can see traffic behind you;	**CUR** Nos 23 and 24
— your horn is in working order;	**CUR** No 29
— your speedometer is in working order;	**CUR** No 98
— your exhaust system is efficient;	**CUR** No 116A
— any audible anti-theft device that may be fitted complies with the regulations;	**CUR** No 29 (6)
— the load on your vehicle is so secured that neither danger nor nuisance is caused by its falling or being blown off, or shifting;	**CUR** No 97 (2)
— your load if it projects sideways or to the front or rear is not of illegal width or length and at night any extra front and rear lamps are carried and are lit;	**CUR** Nos 139 and 140 **RVLR** Nos 18, 20 and 21
— the overall travelling height is recorded in the cab (certain kinds of vehicles only);	**CUR** No 80A

	Act & Regulations
— your vehicle has lamps and reflectors which comply with the regulations and are in working order;	**RVLR** Nos 16 and 20
— your headlamps are properly adjusted;	**RVLR** No 20
— your vehicle, if a private car or other vehicle to which the regulation applies, is fitted with seat belts, which must be maintained in good condition;	**CUR** No 17
— your vehicle, if a road tanker or a vehicle conveying a tank container, carrying a prescribed hazardous substance, displays the required hazard warning panels and that these are kept clean and free from obstruction.	**DS(RTTC)R 1981** No 19

When driving you must

— wear an approved type of seat belt in any vehicle to which the law applies unless you are exempt from so doing;	**RTA 1972** Sect 33A **MV(WSB)R** 1982
— be in such a position that you can exercise proper control over your vehicle and retain a full view of the road and traffic ahead;	**CUR** No 119
— give precedence to a pedestrian who is on an uncontrolled Zebra crossing, that is, a crossing marked by black and white stripes, studs and lighted beacons and at which there is no police constable * controlling the traffic;	**ZPCR** No 8
— give precedence to pedestrians on a Pelican crossing, when an amber light is flashing;	**PPCRGD** No 11
— observe speed limits (70 mph on motorways and dual carriageways and 60 mph on all other roads unless a lower limit is indicated by signs or street lighting) or any special speed limit for your vehicle;	**RTA 1972** Sect 203 **RTRA** Sects 81, 84, 86, 88 and Sch 6 **TA 1982** Sect 61
— observe amber † and red "STOP" signals, traffic signs which give orders, double white lines, yellow road markings and the directions of a police constable * controlling traffic or giving directions for the purposes of a traffic survey;	**RTA 1972** Sect 22 **TSRGD 1981** No 34 (1) **PPCRGD** No 10 **RTRA** Sects 1, 6, 9 and 53 **RTA 1974** Sect 6
— stop when required to do so by a police constable * in uniform;	**RTA 1972** Sect 159
— stop when signalled to do so by a School Crossing Patrol exhibiting a "STOP — CHILDREN" sign;	**RTRA** Sect 28
— see that your side, tail and registration plate lamps are alight at night;	**RVLR** No 21 **RV(R and L)R**
— use your headlamps at night in unlit areas;	**RVLR** No 22
— use your headlamps or front fog lamps in poor daytime visibility.	**RVLR** No 22

You must not

— without reasonable excuse allow a child under 14 to travel in the front of a vehicle to which the law applies unless suitably restrained;	**RTA 1972** Sect 33B **MV(WSBC)R** 1982

** The reference to a police constable includes a traffic warden (Transport Act 1968).*

† See the illustrations on page 46.

	Acts & Regulations
— drive recklessly;	**RTA 1972** Sect 2
— drive without due care and attention or without reasonable consideration for other persons using the road;	**RTA 1972** Sect 3
— on the approach to an uncontrolled Zebra crossing marked by a pattern of zigzag lines, overtake the moving motor vehicle nearest to the crossing or the leading vehicle which has stopped to give way to a pedestrian on the crossing;	**ZPCR** No 10
— drive in a bus lane during its hours of operation,	**RTRA 1967** Sects 1, 6 and 9
— wilfully drive on a footpath by the side of any road made or set apart for the use of foot passengers;	**HA 1835** Sect 72 **RB(S)A** Sch C and **BP(S)A** Sect 381 (13)
— drive under the influence of drink or drugs;	**RTA 1972** Sect 5 (
— drive with a breath alcohol level higher than 35 μg/100ml (equivalent to a blood alcohol level of 80 mg/100ml):	**TA 1981** Sect 25
— drive your vehicle in reverse more than necessary;	**CUR** No 120
— drive a vehicle which emits excessive fumes and smoke;	**CUR** Nos 109, 110, 111 and 112
— drive a vehicle which has an unsuitable or defective silencer;	**CUR** Nos 30, 116A and 116B
— drive a vehicle which emits noise in excess of the maximum limits laid down;	**CUR** No 116 and Sch 9
— sound your horn at night (11.30 p.m.—7 a.m.) in a built-up area;	**CUR** No 118 (1)
— carry passengers in such numbers or in such a manner as is likely to cause danger;	**CUR** No 97 (1)
— use four-way flashing hazard warning lamps on a moving vehicle;	**RVLR** No 23
— switch on rear fog lamps unless visibility is *seriously* reduced;	**RVLR** No 23
— dazzle other roads users with your headlamps or rear fog lamps.	**RVLR** No 23

When you stop you must

— set the brake and stop the engine before you leave the vehicle;	**CUR** Nos 117 and 124
— always switch off your headlamps and, at night, leave your side, tail and registration plate lamps on unless unlit parking is allowed;	**RVLR** Nos 21 and 23
— when required by the police, produce your driving licence, certificate of insurance and, if your vehicle is subject to compulsory testing, your test certificate, for examination. If necessary, you may instead produce them within five days at any police station you select; in prescribed circumstances state your date of birth.	**RTA 1972** Sects 161 and 162

You must not

— stop your vehicle on the approach side of a Pelican crossing beyond the double line of studs in the road except to obey the signals, or in circumstances beyond your control, or when it is necessary to do so to avoid an accident, or to wait to turn right or left;	**PPCRGD** No 9

— stop your vehicle in the Zebra-controlled areas which are marked by a pattern of zigzag lines on either side of an uncontrolled Zebra crossing except to give precedence to a pedestrian on the crossing, or to wait to turn right or left, or in circumstances beyond your control, or when it is necessary to avoid an accident;	**ZPCR** No 12
— stop your vehicle within the limits of any type of pedestrian crossing except in circumstances beyond your control, or to avoid an accident;	**PPCRGD** No 12 **ZPCR** No 9
— park your vehicle or trailer on the road so as to cause unnecessary obstruction;	**CUR** No 122
— park in a bus lane during its hours of operation, except when permitted, to load or unload goods;	**RTRA 1967** Sects 1, 6 and 9
— park your vehicle or trailer on the road in such a way that it is likely to cause danger to other road users;	**RTA 1972** Sect 24
— park your vehicle on any length of road marked with double white lines even if one of the lines is broken;	**RTA 1972** Sect 22, **TSRGD 1981** No 23 (2) (a)
— park on any verge, central reservation or footway, if your vehicle is a goods vehicle with a maximum laden weight exceeding 7·5 tonnes, except in certain circumstances, for example, if loading or unloading could only be performed there and the vehicle is not left unattended;	**RTA 1972** Sect 36A **HCVA 1973** Sect 2
— park at night on the right-hand side of the road (except in a one-way street);	**CUR** No 123
— park your vehicle at night without lights unless: the road is subject to a speed limit of 30 mph or less; your vehicle is parked with its near side close to the kerb (unless it is in a one-way street or a recognised parking place); and no part of your vehicle is within 15 yards of a road junction;	**RVLR** No 21
— park your vehicle contrary to waiting restrictions; on a clear-way you must not stop your vehicle on the carriageway; in a parking meter zone you must not park except at a meter; and outside meter zone hours you must not park except in conformity with local waiting restrictions or a bus lane order;	**RTRA** Sects 1, 6, 9 and 53
— park your vehicle on common land more than 15 yards from a highway;	**RTA 1972** Sect 36
— sound your horn while stationary, except in times of danger due to another moving vehicle nearby;	**CUR** No 118 (1)
— open any door of your vehicle so as to cause injury or danger to anyone.	**CUR** No 125

If you are involved in an accident

— which causes damage or injury to any other person, or other vehicle, or any animal (horse, cattle, ass, mule, sheep, pig, goat or dog) not in your vehicle, or roadside property:	**RTA 1972** Sects 25 and 166 **RTA 1974** Sch 6

You must

(a) stop;

(b) give your own and the vehicle owner's name and address and the registration mark of the vehicle to anyone having reasonable grounds for requiring them;

(c) if you do not give your name and address to any such person at the time, report the accident to the police as soon as reasonably practicable, and in any case within 24 hours;

(d) if anyone is injured and you do not produce your certificate of insurance at the time to the police or to anyone who has with reasonable grounds required its production, report the accident to the police as soon as possible, and in any case within 24 hours, and either produce your certificate of insurance to the police when reporting the accident or ensure that it is produced within five days thereafter at any police station you select.

To drivers and front seat passengers

You must

— wear an approved type of seat belt in any vehicle to which the law applies unless you are exempt from so doing. — **RTA 1972** Sect 33, **MV(WSB)R** 1982

To motorcyclists and moped riders

Most of the requirements of the law relating to motor drivers, including those relating to pedestrian crossings, apply to you. In addition:

You must

— wear an approved type of safety helmet on all journeys. — **MC(PH)R**

— ensure that your exhaust system and silencer are of an approved type. — **CVR** 116C

You must not

— carry more than one passenger on a a two-wheeled machine, and the passenger must sit astride the cycle on a proper seat securely fitted behind the driver's seat and with proper rests for the feet; — **RTA 1972** Sect 16 and **CUR** No 141

— park in a parking meter zone except in a specially marked motorcycle park or, where not prohibited by a local order, at a meter. — **RTRA** Sect 53

To pillion passengers

You must

— wear an approved type of safety helmet whenever you ride as pillion passenger. — **MC(PH)R**

Motorway driving

General

	Acts & Regulations
Pedestrians, learner drivers, pedal cycles, motorcycles under 50 cc capacity, invalid carriages not exceeding 5 cwt unladen weight, certain slow-moving vehicles carrying oversized loads (except by special permission), agricultural vehicles and animals must not use motorways.	**HA 1980** Sects 16 and 17 **SRCTO** and **MT(E&W)R** No 11, **SRA** Sects 1 and 2; **SRCT(S)O** and **MT(S)R** No 10

To drivers and motorcyclists

You must

	Acts & Regulations
— drive on the carriageways only;	**MT(E&W)R** No 5, **MT(S)R** No 4
— observe one-way driving on the carriageways;	**MT(E&W)R** No 6, **MT(S)R** No 5
— keep any animals in your charge in the vehicle or (in emergency) under proper control on the verge;	**MT(E&W)R** No 14 **MT(S)R** No 12
— observe speed limits or any special speed for your vehicle;	**RTRA** Sect 17 and **RTRA** Sect 86 and Sch 6
— observe flashing red signals when displayed over your lane or when displayed at the side of the carriageway (for example, on a slip road).	**RTA 1972** Sect 22, **TSRGD 1981** No 34 (4)

You must not

	Acts & Regulations
— use a motorway if you are a learner driver (except heavy goods vehicle drivers);	**MT(E&W)R** No 11 **MT(S)R** No 10
— exceed 70 mph at any time;	**MT(SL)R**
— reverse on the carriageways;	**MT(E&W)R** No 8, **MT(S)R** No 7
— stop on the carriageways;	**MT(E&W)R** No 7(1) **MT(S)R** No 6(1)
— stop on the hard shoulder except in an emergency;	**MT(E&W)R** No 9 **MT(S)R** No 8
— stop on the central reservation or verge;	**MT(E&W)R** No 10 **MT(S)R** No 9
— walk on the carriageway or on the central reservation except in an emergency;	**MT(E&W)R** No 13 **MT(S)R** No 11
— if you are driving a goods vehicle which has an operating weight of more than 7·5 tonnes, or any vehicle drawing a trailer, or a bus longer than 12 metres — use the right-hand lane of a motorway with three or more lanes except in exceptional circumstances as prescribed.	**MT(E&W)R** No 12 **MT(E&W)(A)R** **MT(S)R** No 10A

First aid on the road

For those with no first aid training.

DANGER — Deal with threatened danger or you and the casualties may be killed. FURTHER COLLISIONS and FIRE are the dangers in a road accident.

Action: If possible warn other traffic. Switch off the engine. Impose a "No Smoking" ban.

Obtaining further help: Send a bystander to call an ambulance as soon as possible; state the exact location of accident and the numbers of vehicles and casualties involved.

People remaining in vehicles: Casualties remaining in vehicles should not be moved unless further danger is threatened.

If breathing has stopped: Remove any obvious obstruction in the mouth. Keep the head tilted backwards as far as possible — breathing may begin and the colour may improve. If not, pinch the casualty's nostrils together and blow into the mouth until the chest rises; withdraw, then repeat regularly once every four seconds until the casualty can breathe unaided.

If unconscious and breathing: Movement may further damage an injured back, so only move if in danger. If breathing becomes difficult or stops, treat as in foregoing paragraph.

If bleeding is present: Apply firm hand pressure over the wound, preferably using some clean material without pressing on any foreign body in the wound. Secure a pad with a bandage or length of cloth. Raise limb to lessen the bleeding, providing it is not broken.

Reassurance: The casualty may be shocked but prompt treatment will minimise this; reassure him confidently; avoid unnecessary movement; keep him comfortable and prevent him getting cold; ensure he is not left alone.

Give the casualty NOTHING to drink.

Carry a first aid kit. Learn first aid — from the St John Ambulance Association and Brigade, St Andrew's Ambulance Association or the British Red Cross Society.